RAND NATIONAL SECURITY RESEARCH DIVISION

Improving Federal and Department of Defense Use of Service-Disabled Veteran-Owned Businesses

Amy G. Cox, Nancy Y. Moore

Prepared for the Office of the Secretary of Defense

Approved for public release; distribution unlimited

The research described in this report was prepared for the Office of the Secretary of Defense (OSD). The research was conducted within the RAND National Defense Research Institute, a federally funded research and development center sponsored by OSD, the Joint Staff, the Unified Combatant Commands, the Navy, the Marine Corps, the defense agencies, and the defense Intelligence Community under Contract W74V8H-06-0002.

Library of Congress Cataloging-in-Publication Data

Cox, Amy G.
  Improving federal and Department of Defense use of service-disabled veteran-owned businesses / Amy G. Cox, Nancy Y. Moore.
    pages cm
  Includes bibliographical references.
    1. Business enterprises owned by veterans with disabilities—Government policy—United States. 2. Veteran-owned business enterprises—Government policy—United States. 3. Public contracts—Government policy—United States. 4. United States. Department of Defense—Rules and practice. I. Moore, Nancy Y., 1947- II. Title.

    UB357.C69 2013
    352.5'3—dc23

                                                                    2013027174

The RAND Corporation is a nonprofit institution that helps improve policy and decisionmaking through research and analysis. RAND's publications do not necessarily reflect the opinions of its research clients and sponsors.

Support RAND—make a tax-deductible charitable contribution at www.rand.org/giving/contribute.html

RAND® is a registered trademark.

© Copyright 2013 RAND Corporation

This document and trademark(s) contained herein are protected by law. This representation of RAND intellectual property is provided for noncommercial use only. Unauthorized posting of RAND documents to a non-RAND website is prohibited. RAND documents are protected under copyright law. Permission is given to duplicate this document for personal use only, as long as it is unaltered and complete. Permission is required from RAND to reproduce, or reuse in another form, any of our research documents for commercial use. For information on reprint and linking permissions, please see the RAND permissions page (www.rand.org/pubs/permissions.html).

RAND OFFICES
SANTA MONICA, CA • WASHINGTON, DC
PITTSBURGH, PA • NEW ORLEANS, LA • JACKSON, MS • BOSTON, MA
DOHA, QA • CAMBRIDGE, UK • BRUSSELS, BE
www.rand.org

Preface

The federal government has several policies to boost economic opportunities for service-disabled veterans. Among these is a goal to award 3 percent of its prime-contract procurement dollars to service-disabled veteran-owned small businesses (SDVOSBs). Federal agencies also have a goal to award 3 percent of subcontracting dollars to SDVOSBs. Nevertheless, federal contracting with SDVOSBs remains below this 3 percent goal, although it did increase from 1.4 percent in fiscal year (FY) 2008 to 2.7 percent in FY 2011. Similarly, the Department of Defense (DoD) has increased the number of prime-contract dollars going to SDVOSBs but has not yet reached the government-wide goal, awarding 2.0 percent of its prime-contract awards to SDVOSBs in FY 2009.[1]

As part of its commitment to this effort, DoD asked the RAND Corporation to investigate the SDVOSB procurement program. In particular, DoD's Office of Small Business Programs (OSBP) asked RAND to identify barriers that SDVOSBs, the federal government, and DoD face in increasing the percentage of prime-contract dollars going to SDVOSBs. This report is the result of that research. In it, we assess the characteristics of service-disabled veterans that support successful business ownership, identify barriers that DoD and other federal contracting personnel have in meeting their SDVOSB prime-contract goals, and identify barriers SDVOSBs face when seeking DoD and other federal prime contracts and subcontracts. This research should be of interest to persons or organizations interested in small business, service-disabled veterans, and procurement, such as the Interagency Task Force for Veterans Small Business Development.

This research was sponsored by the DoD Office of Small Business Programs and was conducted within the Acquisition and Technology Policy Center of the RAND National Defense Research Institute, a federally funded research and development center sponsored by the Office of the Secretary of Defense, the Joint Staff, the Unified Combatant Commands, the Navy, the Marine Corps, the defense agencies, and the defense Intelligence Community.

For more information on the RAND Acquisition and Technology Policy Center, see http://www.rand.org/nsrd/ndri/centers/atp.html or contact the director (contact information is provided on the web page).

---

[1]   On small-business procurement data, see Federal Procurement Data System—Next Generation, "Small Business Goaling Reports," undated.

Contents

Preface iii

Figure and Tables vii

Summary ix

Acknowledgments xv

Abbreviations xvii

CHAPTER ONE

Characteristics of Veterans and Service-Disabled Veterans and Their Potential Effect on the Supply of SDVOSBs 1

Introduction 1

Individual Characteristics Analyses 1

CHAPTER TWO

Interview Approach and Results 7

Strategy for Sampling Industries 7

Interview Methodology 8

    Strategy for Sampling SDVOSBs and Contract Staff in the Chosen Industries 8

    Interview Operations 9

Interview Results 11

    Barriers Faced Particularly by SDVOSBs 12

    Barriers Also Faced by Non-Traditional Suppliers and Small Businesses 14

CHAPTER THREE

Summary of Findings and Recommendations 17

Increase the Priority of the SDVOSB Program 18

Strengthen Education and Training Opportunities 18

Continue Efforts to Remove Barriers to Non-Traditional Suppliers and Other Small Businesses 19

APPENDIXES

A. Logistic Regression Methodology and Analyses 21

B. Interview Protocols 25

Bibliography 29

Figure and Tables

Figure

1.1. Number of Veterans and Service-Disabled Veterans, 1986 to 2011 4

Tables

1.1. Population Estimates and Paid Labor Force Status of the Adult Population, by Veteran and Service-Disability Status 2

1.2. Education Level and Age of Adult Population, by Veteran and Service-Disability Status 3

1.3. Odds of Self-Employment, by Veteran and Service-Disability Status and Paid Labor Force Participation, Controlling for Other Factors 4

1.4. Percentage of Veterans Self-Employed and Disabled, by Cohort 5

1.5. Average Annual Business Revenue, by Veteran Status 6

2.1. Characteristics of SDVOSBs Sampled and SDVOSBs Interviewed 11

A.1. Odds Ratios from Logistic Regression of Self-Employment Among All Adults 22

A.2. Odds Ratios from Logistic Regression of Self-Employment Among All Adults in Paid Labor Force 23

A.3. Logistic Regression Results of Self-Employment, by Veteran and Disability Status 24

Summary

The federal government's goal of awarding 3 percent of all prime-contract dollars to service-disabled veteran-owned small businesses (SDVOSBs) (Bush, 2004; U.S. Congress, 2003, 1999) has not yet been achieved. The objective of this report is to identify barriers to federal contracting with such firms. It summarizes findings and presents recommendations from analyses of quantitative data on service-disabled veterans and of interviews with SDVOSBs and the federal contracting personnel who work with them.

Quantitative Analyses

Using person-level data on veterans, service-disabled veterans, and non-veterans, we consider individual-level barriers that might reduce the likelihood that they will own businesses.

We found that service-disabled veterans are less likely than both veterans and non-veterans to be self-employed. Service-disabled veterans' lower self-employment does not appear to be explained by a lack of education or experience in the labor market, because, on average, service-disabled veterans have more education than non-veterans and all veterans and are older than non-veterans. Even when differences in education, age, gender, race-ethnicity, and marital status are controlled for simultaneously, service-disabled veterans are still less likely than non-disabled veterans and less likely than non-veterans to be self-employed. Service-disabled veterans' odds of self-employment were 33 percent lower than for non-veterans and 48 percent lower than for non-disabled veterans—differences that are statistically significant.

Recent apparent increases in the number of service-disabled veterans appear unlikely to translate into increases in the number of SDVOSBs with whom the federal government might contract. This is because most recent veterans appear to have higher levels of disability and lower rates of self-employment than veterans from earlier cohorts.

Interviews

A central focus of this project is the identification of barriers experienced by SDVOSBs when trying to contract with the federal government. To assess these barriers, we conducted in-depth interviews with 13 SDVOSBs. We also interviewed DoD contracting staff who had worked with SDVOSBs to learn about their experiences and to obtain a more complete understanding of the barriers SDVOSBs face.

Methodology

We used five criteria to select firms for interviews. Specifically, we sought SDVOSBs in industries

1. with a sufficient number of SDVOSBs to provide for a sufficient number of responses
2. with sufficient federal spending to enable contracting with the federal government
3. where SDVOSBs have won some contracts
4. where SDVOSBs represent a significant portion of registrants in the Central Contractor Registration (CCR), that is, where federal contracting was not precluded by too few SDVOSBs
5. where SDVOSBs might have competition from businesses qualifying for other preference programs.

We used tabulations of 2011 Federal Procurement Data System contract-action data and CCR data to select the industries where we would focus our interviews. We used the North American Industry Classification System (NAICS) to define industries, and we ranked industries by the number and percentage of SDVOSBs registered in them, the amount of total federal spending in them, and the amount of federal spending awarded to SDVOSBs. We subsequently identified Engineering Services (NAICS number 541330), Commercial and Institutional Building Construction (NAICS number 236220), and Facilities Support Services (NAICS number 561210) as the three industries on which to focus our interviews.

We then sampled SDVOSBs in these industries to interview. We randomly selected firms within each of these three industries and within each of 12 groups defined by funding source, other preference-program membership, and Small Business Administration certification. We sampled 132 companies to solicit for participation in the study. We also randomly selected 20 contracting specialists in each of the three industries from all DoD contracts with SDVOSBs, for a total of 60 contracting specialists to solicit for participation. We developed interview protocols from the literature and policy presented in a background paper prepared by the Office of Small Business Programs (OSBP) at DoD.

We received responses from 13 companies and five contract staff. We rigorously analyzed the responses for themes and for insight into the barriers to federal contracting that SDVOSBs may face.

Results

The SDVOSBs interviewed were similar to those sampled with regard to their qualification for other preference programs, their size, and, to a lesser degree, their federal contracting experience. The interviewed firms also shared a broad range of other characteristics, including years in business, ownership of other businesses, proportion of commercial and federal contracts, and number and amount of federal contract bids and awards. The firms interviewed had worked with 21 federal agencies, the most common of which were the Department of Veterans Affairs (VA) and DoD.

The interviews identified three barriers unique to these firms because they were SDVOSBs: the prior definition of the SDVOSB contracting program as a less-urgent goal than other small business programs, a lack of understanding among many SDVOSBs interviewed about the program, and a barrier to receiving planned subcontracts.

The first barrier is that the SDVOSB contracting program's goal was first defined with less-rigorous language than the goals of other programs. The specific language in the Federal Acquisition Regulation (FAR) for SDVOSBs had, until a recent revision, used the term "may set aside acquisitions" for this group, whereas the language used for other preferred businesses, such as those in historically underutilized business zones, was "shall set aside acquisitions." This distinction of "may" versus "shall" was recently erased by the 2010 Small Business Jobs Act, which established parity among the different programs (U.S. Congress, 2010). When incorporating this change into the FAR, program administrators noted their intention to clarify "that there is no order of precedence among the small business socioeconomic contracting programs" (*Federal Register*, 2012). Nevertheless, the distinction has historically meant that contracting staff had greater discretion on whether to set aside or award a contract to an SDVOSB. Many SDVOSBs and some contract staff interviewed maintained that this discretion continues to result in a lower priority for SDVOSB contracting. Some contracting staff reported that the lower priority of SDVOSB contracting is exacerbated by the fact that goals for other small business procurement programs draw the efforts of contracting staff because the other goals are higher. The lack of knowledge and prior experience in federal contracting of many SDVOSBs interviewed, requiring that contracting staff and customers do more work and absorb more risk in working with such businesses, could further reduce the likelihood that an SDVOSB will receive a contract.

A second but related barrier described by these interviewees—and one that is unique to SDVOSBs—is that many did not understand that the SDVOSB procurement goals are not mandatory and are not guaranteed. We heard many misconceptions about the program in the SDVOSB interviews, despite the fact that the interviewees had attended federal information seminars about the program.

Finally, for prime contractors, including an SDVOSB on a bid demonstrates an effort to satisfy federal SDVOSB subcontracting goals, but a lack of federal oversight after awards are made reportedly fails to ensure that prime contractors include SDVOSBs in the actual work. Planning to use SDVOSBs for subcontracting to meet federal subcontracting goals improves a company's chances of winning a contract. Many SDVOSB respondents described experiences with prime contractors that included them as subcontractors on bids but then rarely included them in awarded work. These SDVOSBs were reluctant to complain about this problem to federal officials for fear of being labeled as whistle-blowers and being excluded from future subcontracts. The effect of making complaints could extend beyond the lost subcontracting business into future prime contracts as well.

The interviewees also described encountering many barriers faced by non-traditional suppliers and other small businesses as well. These included the inherent complexities of the FAR and of the federal bidding process, a lack of sufficient federal educational and networking opportunities, a lack of communication from key contracting personnel, the slow federal processes for award decisions and making final payments, insufficient knowledge among SDVOSBs about their chances of winning a bid, the risk of wasting resources on developing a bid that was likely to go to an incumbent or another established supplier or of investing in a bid that was subsequently cancelled, and the possibility of being in an emerging industry that does not fit existing NAICS categories.

The SDVOSBs also described some limited sources of support for managing these barriers. These included the federal training conferences, which provide very basic information, and more detailed information and assistance from state-level Procurement Technical Assistance Centers (PTACs). A few SDVOSBs described partnering with other suppliers on bids as the most useful source of information. Firms that reported partnering with other suppliers or hiring staff with prior federal contracting knowledge improved their bidding further. Finally, SDVOSBs that had a strong customer-oriented approach to federal contracting described the greatest success.

For the SDVOSBs interviewed that were either unaware of or unable to invest in these sources of support, the barriers often created some level of disillusionment with the program and with the federal government's commitment to veterans. The widely publicized 3 percent goal raised the expectations of many of these SDVOSBs that they would receive federal business, and they invested in their companies and in bids accordingly. However, their experience was facing fewer bidding opportunities, greater competition, and fewer winning bids than they expected. Exceptions to this disillusionment were those few SDVOSBs interviewed that focused relentlessly on increasing their knowledge of the FAR and of the bidding process and that catered to the needs of both the contracting officer and the customer when they did win an award.

Our findings suggest multiple reasons for the low rate of federal SDVOSB contracting compared with the 3 percent federal goal. First, the level of priority placed on the program likely contributes to the federal government's inability to meet its goal. The policy has only very recently been worded as strongly as that for other small business procurement programs; competing goals from other programs limit set-asides for SDVOSBs, and limited oversight of subcontracting goals may discourage prime contractors from including SDVOSBs in awarded work. Second, SDVOSBs, especially new ones, may also confront the same barriers faced by small businesses and non-traditional suppliers. Third, although some sources of support for managing these barriers exist, they are limited. Finally, given these barriers, many SDVOSBs interviewed were disillusioned with the program, and, to the extent that such disillusionment is widespread, it may limit federal contracting rates further.

Recommendations

Given these barriers, whether the federal government can reach its 3 percent goal for SDVOSB contracting is not clear. Identifying the number of SDVOSBs in the industries in which federal agencies spend, particularly by federal agency, may be useful, as would analyses of service-disabled veterans' decisionmaking processes regarding self-employment and estimations of the actual dollar amounts and number of contracts that would be received by SDVOSBs. In addition, reducing barriers facing SDVOSBs may involve increasing the priority of the SDVOSB program, strengthening education and training opportunities, and continuing efforts to remove barriers to non-traditional suppliers and small businesses.

Increase the Priority of the SDVOSB Program

If the federal government wants to increase the priority of the SDVOSB program, federal agencies need to ensure that the recent changes to the policy made by the 2010 Small Business Jobs Act are transmitted explicitly to federal contracting staff. Agencies also need to inform staff of which industries have ample numbers of SDVOSBs, especially as compared with the numbers of firms in other preference programs. A final step toward raising the priority of the SDVOSB program would be to provide contracting resources for reviewing prime contractors' execution of their bid and subcontracting plans, conduct those reviews, and publish the results.

Strengthen Education and Training Opportunities

Providing SDVOSBs also with the industry analyses described above could be a first step toward strengthening their education and training opportunities. Additional education could include more advanced information and training about the FAR and the federal bidding process and about the roles and responsibilities of contracting staff. Strengthening and enhancing PTAC support of these efforts could also strengthen education and training opportunities and may be most efficient. Networking opportunities in any of these forums could help SDVOSBs identify firms with which to partner or subcontract to learn more about the bidding process.

Finally, clarifying federal expectations of SDVOSBs might also help strengthen education and training opportunities. Some SDVOSBs interviewed saw dramatic improvements in their ability to win contracts when they invested effort in understanding the FAR and in the lengthy process of learning how to develop a successful bid. Similarly, the SDVOSBs interviewed that expected to have to learn and meet the needs of both the contracting officer and the customer reportedly received strong performance evaluations, which strengthened their ability to win future bids.

Continue Efforts to Remove Barriers to Non-Traditional Suppliers and Other Small Businesses

Barriers faced, at least by the SDVOSBs we interviewed, can also be reduced by more general efforts to remove barriers to non-traditional suppliers and other small businesses. These efforts include continuing to streamline the administrative requirements of and removing inefficiencies from the bidding process, further streamlining the award decision and closing payment procedures, improving communication between government and suppliers, and helping suppliers in emerging industries become known to both customers and contracting officers.

Acknowledgments

This research would not have been possible without the contributions of many people. First and foremost, we thank the service-disabled veteran-owned small businesses (SDVOSBs) and the federal contracting staff members who generously shared their time and experience for this research. Without their contributions, we would have only a shallow understanding of the barriers facing the SDVOSB procurement program. We also wish to extend deep thanks to our sponsor, the Office of Small Business Programs at the Department of Defense, for its support of this work. In particular, Andre Gudger, Linda Oliver, Paul Simpkins, Linda Robinson, Rick Rhoadarmer, Don Gilbert, and Maggie Sizer all made notable contributions to this work. We also thank our RAND colleagues for their excellent work assisting this project: Cynthia Cook, Paul DeLuca, Clifford Grammich, Judith Mele, Patricia Bedrosian, and Donna Mead, and the report's reviewers, John Ausink and John Winkler.

Abbreviations

CCR	Central Contractor Registration
CPS	Current Population Survey
DoD	Department of Defense
FAR	Federal Acquisition Regulation
FY	fiscal year
NAICS	North American Industry Classification System
OSBP	Office of Small Business Programs
PTAC	Procurement Technical Assistance Center
RFP	request for proposal
SBA	Small Business Administration
SDVOSB	service-disabled veteran-owned small business
VA	Department of Veterans Affairs

Characteristics of Veterans and Service-Disabled Veterans and Their Potential Effect on the Supply of SDVOSBs

Introduction

The federal government has a goal of awarding 3 percent of all prime-contract dollars to service-disabled veteran-owned small businesses (SDVOSBs), and federal agencies have an additional goal that 3 percent of its subcontracting dollars go to SDVOSBs as well (Bush, 2004; U.S. Congress, 2003, 1999; Department of Defense [DoD], 2007). Neither the federal government nor DoD has met its goal. The government's percentage did increase from 1.39 percent to 2.65 percent between fiscal year (FY) 2008 and FY 2011, and DoD's percentage increased from 0.94 in FY 2008 to 2.02 in FY 2011 (Federal Procurement Data System—Next Generation, undated). This report identifies barriers that might be preventing the government from achieving its goals. It summarizes findings and presents recommendations from analyses of quantitative data on service-disabled veterans and of interviews with SDVOSBs and federal contracting personnel who work with SDVOSBs. The objective of these analyses is to identify barriers to federal contracting with such firms.

We first provide a quantitative analysis of veterans to identify any individual characteristics that could be associated with barriers to achieving the goals. In Chapter Two, we describe our interview methods and results, and in Chapter Three, we conclude with a summary of our findings and recommendations.

Individual Characteristics Analyses

In this section, we explore person-level data on veterans and service-disabled veterans to determine whether individual characteristics that could be associated with barriers might limit the likelihood that they will own such businesses.

We use the Current Population Survey (CPS) (U.S. Bureau of Labor Statistics, 2010)—the federal survey used to calculate monthly unemployment rates and other national economic figures—for these quantitative analyses. The July 2010 CPS included a supplemental survey of veterans, which we tabulated. We restricted the analysis to adults over age 18. We define veterans as anyone with a history of military service and service-disabled veterans as anyone with a service-connected disability.

An initial tabulation of these data show that service-disabled veterans are less likely than both veterans and non-veterans to be self-employed (Table 1.1). Whereas 7 percent of non-veterans and veterans are self-employed, only 4 percent of service-disabled veterans are. This difference is statistically significant at the 0.001 level.

Table 1.1
Population Estimates and Paid Labor Force Status of the Adult Population, by Veteran and Service-Disability Status

	Non-Veterans	All Veterans	Service-Disabled Veterans
Total estimated in population over age 18	205,000,000	20,000,000 (9% of adults)	2,800,000 (15% of veterans)
Employment status			
Private sector employed	47%	31%	26%
Government-employed	9%	11%	16%
Self-employed	7%	7%	4%
Unemployed	6%	4%	5%
Retired	14%	37%	31%
Disabled, unable to work	5%	6%	13%
Other, not employed	12%	4%	5%
Sample size	90,687	9,736	1,235

SOURCE: Authors' tabulation of July 2010 CPS data.

NOTES: Because the CPS captures only the civilian, non-institutionalized population, these totals are likely underestimates. They do not include persons who are homeless or living in prisons, hospitals, or other institutions.

Why might service-disabled veterans be less likely to own their own businesses? Previous research on small business owners identified three factors as key to small business success: access to capital, a family history with successful small business, and education (Fairlie and Robb, 2008). Service-disabled veterans do have at least some access to both social and financial capital for small businesses through government programs designed specifically to assist veterans and service-disabled veterans with starting and maintaining a small business (see, for example, U.S. Small Business Administration, 2012a, 2012b). Unfortunately, we do not have information on the family histories of service-disabled veterans and thus cannot determine whether their families' rates of business ownership is any different from that of other veterans or non-veterans. However, we do have information on the education level of service-disabled veterans.

Table 1.2 presents the education levels of non-veterans, veterans, and service-disabled veterans. It shows that, on average, service-disabled veterans have more education than non-veterans and all veterans. Service-disabled veterans are less likely than non-veterans and veterans to have only a high school education or some college (a statistically significant difference), whereas all three groups have similar rates of earning bachelor's degrees and graduate degrees. Thus, a lack of education does not appear to be the reason behind service-disabled veterans' lower rate of self-employment.

As with education, age can be positively associated with business success because it is a proxy for time in the paid labor force. However, lack of experience (as suggested by age) does not appear to explain service-disabled veterans' lower rates of business ownership either. Table 1.2 shows the average age of the three groups and also shows that veterans and service-disabled veterans are older than the non-veteran population (a statistically significant differ-

Table 1.2
Education Level and Age of Adult Population, by Veteran and Service-Disability Status

	Non-Veterans	All Veterans	Service-Disabled Veterans
Education			
High school graduate or less	43%	39%	32%
Some college	28%	34%	40%
Bachelor's degree	19%	17%	17%
Graduate degree	10%	10%	11%
Mean age (age 19–85)	46 years	59 years	56 years
Sample size	90,687	9,736	1,235

SOURCE: Authors' tabulation of July 2010 CPS data.

ence). Veterans are older because they are not evenly distributed across cohorts; larger numbers served in the Vietnam War than in later years and conflicts. As noted, greater age usually translates into more employment experience and therefore tends to be associated with stronger labor market positions, such as higher rates of self-employment. At the same time, developing a business to the point where it can be a successful government contractor usually takes many years, and persons who are older may not have enough working years left before retirement to devote to this long process. Although age does not seem to lower self-employment among veterans as a whole, perhaps service-disabled veterans are less likely to start a business at older ages.

One way to add more clarity to these results is to use a multivariate analysis, which examines the potential relationship of self-employment to multiple characteristics at the same time. We used logistic regression to examine whether service-disabled veterans are still less likely to be self-employed when we control for education, age, and other differences between them and other veterans and between them and non-veterans. We found that even when differences in education, age, gender, race-ethnicity, and marital status are controlled for, service-disabled veterans are still less likely than non-disabled veterans and less likely than non-veterans to be self-employed. Table 1.3 shows that service-disabled veterans' odds of self-employment were 33 percent lower than that of non-veterans and 48 percent lower than that of non-disabled veterans. These differences are statistically significant at the $p < 0.001$ level.

In addition, when we restricted the analysis to only those already in the paid labor force (i.e., we excluded those who are retired or are out of the paid labor force for other reasons), these differences only increased. Among adults in the paid labor force, service-disabled veterans' odds of self-employment are 43 percent lower than non-veterans' odds and 63 percent lower than non-disabled veterans' odds. These differences are also statistically significant at the $p < 0.001$ level. Detailed results for these and related analyses can be found in Appendix A.

The multivariate analyses did not identify any individual-level barriers to business ownership faced by service-disabled veterans. However, recent increases in the number of service-disabled veterans, or at least in the number identified and certified, raise the possibility that an increase in the number of SDVOSBs might occur simply through a population increase (rather than by identifying and lowering barriers). The Department of Veterans' Affairs (VA) tabulates annual estimates of the number of veterans and service-disabled veterans. Because

Table 1.3
Odds of Self-Employment, by Veteran and Service-Disability Status and Paid Labor Force Participation, Controlling for Other Factors

	All Adults	Adults in Paid Labor Force
Service-disabled veterans compared to non-veterans	33% lower odds	43% lower odds
Service-disabled veterans compared to non-disabled veterans	48% lower odds	63% lower odds
Non-disabled veterans compared to non-veterans	69% lower odds	69% lower odds

SOURCE: Authors' tabulation of July 2010 CPS data.

the VA uses a different methodology than the CPS, the VA numbers are slightly higher. These differences include VA attempts to estimate the number of all veterans, whereas the CPS tallies only those who are not homeless, not living in institutions (e.g., hospitals, nursing homes, jails, and prisons), and not living outside the United States. The Department of Veterans' Affairs (2012) notes that between 1986 and 2011, the number of U.S. veterans fell by seven million,[1] whereas the number of service-disabled veterans increased by nearly one million, shown in Figure 1.1. If we assume that the percentage of service-disabled veterans who start businesses is relatively constant, then we would expect that the identified increase in the number of service-

Figure 1.1
Number of Veterans and Service-Disabled Veterans, 1986 to 2011

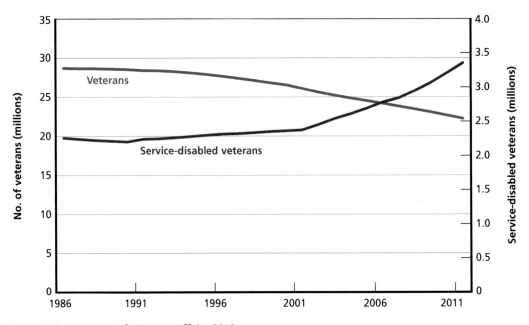

SOURCE: Department of Veterans Affairs, 2012.
RAND RR322-1.1

---

[1]   The decrease in the number of veterans is due in part to the mortality of a very large number of World War II veterans and the decreasing size of the active duty military. In addition, between 1980 and 2010, veterans have increasingly concentrated in smaller, more rural and non-metropolitan counties, often near active military bases, the number of which has also shrunk as a result of base closures and realignments (Teachman, 2013).

disabled veterans from 1986 to 2011 might lead to a corresponding increase in the number of SDVOSBs.

However, the increase in the number of service-disabled veterans has not been uniform but has been concentrated among the more severely disabled. Overall, the number of service-disabled veterans increased between 1986 and 2011, but the number of veterans with low-level disabilities (0 to 20 percent disabled) remained about the same (Department of Veterans' Affairs, 2012). The number of veterans with more severe disabilities appears to have increased substantially. Between 1986 and 2010, the number of veterans with a 30 to 40 percent disability increased from 500,000 to 700,000 (a 40 percent increase); the number of veterans with a 50 to 60 percent disability increased from 200,000 to 500,000 (a 150 percent increase); and the number with the most severe disability rating, 70 to 100 percent disabled, tripled, from less than 300,000 to over 900,000 (Department of Veterans' Affairs, 2012). We have seen that business ownership is less likely among service-disabled veterans than among non-disabled veterans. If, in the same way, business ownership is less likely the more severe a veteran's disability is, then the recent increase in the number of service-disabled veterans, who are disproportionately more severely disabled, is not likely to be accompanied by an increase in SDVOSBs.

Tabulations of the CPS data further suggest that this may be the case. As shown in Table 1.4, more recent veterans have both higher rates of service-connected disability and lower rates of self-employment. In 2010, 15 percent of veterans who served between 1975 and 1990 were service-disabled, compared with 22 percent of veterans who served between 1990 and 2001 and with 27 percent of those who served in 2001 or later. Higher rates of disability among more recent veterans could be the result of multiple factors, including combat survival rates that have increased over time; an increase in the average number of medical conditions in each veteran's claim (Brewin, 2012); the possibility that disabled veterans, especially those from earlier generations, have shorter life spans; and the possibility that disability is less stigmatized, leading veterans to report it more. At the same time, self-employment rates are lower among more recent veterans, dropping from 8 percent for the earlier cohorts of veterans to 3 percent among the most recent veterans. It is not clear whether self-employment among the newest veterans will increase as they gain more experience in the labor market. The 2010 rates shown below suggests that their self-employment would have to more than double to reach the rates of earlier groups of veterans. Together with the VA data, the data shown in Table 1.4 suggest that increased federal contracting with SDVOSBs is unlikely to come simply from the recent increase in the number of service-disabled veterans.

Finally, the revenue of businesses owned by all veterans suggests that, whether or not the number of SDVOSBs increased, successful government contracting involves much more than

Table 1.4
Percentage of Veterans Self-Employed and Disabled, by Cohort

	1975–1990	1990–2001	2001 or later
Percentage of those in cohort service-disabled	15	22	27
Percentage self-employed, among those in paid labor force	8	6	3
Sample size	1,803	982	623

SOURCE: Authors' tabulation of July 2010 CPS data.

simply the numbers of firms. Tabulations of the U.S. Census Bureau's 2007 *Survey of Business Owners* suggest that veteran-owned businesses have substantially less revenue than other businesses. Annual revenue for all businesses averaged more than $1.1 million in 2007; that for veteran-owned businesses was only $498,260 (see Table 1.5). We consider it likely that SDVOSBs, as with other veteran-owned businesses, also have lower revenues than other businesses. If more successful businesses, as indicated by higher revenues, are more likely to obtain government contracts, then this might also contribute to the lower representation of SDVOSBs on government contracts.

Together, these analyses suggest that service-disabled veterans have lower rates of business ownership than both other veterans and non-veterans and that these lower rates are not explained by differences in their education, age, or other background factors. In addition, although the number of service-disabled veterans appears to have increased in recent years, this is unlikely to translate into a significant increase in the development of new and successful SDVOSBs with which the federal government might contract.

Table 1.5
Average Annual Business Revenue, by Veteran Status

	All Business Owners	All Veterans	Service-Disabled Veterans
Average revenue	$1,108,460	$498,260	N/A

SOURCE: Authors' tabulation of U.S. Census Bureau, 2007.

CHAPTER TWO

Interview Approach and Results

A central focus of this project is the identification of barriers experienced by SDVOSBs when trying to contract with the federal government. To assess these barriers, we interviewed several SDVOSBs. We also interviewed contracting staff at DoD who have worked with SDVOSBs to learn about their experiences and to obtain a more complete understanding of the barriers SDVOSBs may face.

Strategy for Sampling Industries

The goal of the study is to find ways to increase federal and DoD contracting with SDVOSBs. To this end, we sought to interview SDVOSBs representative in industries

1. with a sufficient number of SDVOSBs to provide for a sufficient number of responses
2. with sufficient federal spending to enable contracting with the federal government
3. where SDVOSBs have won some contracts
4. where SDVOSBs represent a significant portion of registrants in the Central Contractor Registration (CCR), that is, where federal contracts was not precluded by too few SDVOSBs
5. where SDVOSBs might have competition from businesses qualifying for other preference programs.

We used data generated by the Office of Small Business Programs (OSBP) for FY 2011 Federal Procurement Data System contract-action data and CCR data to select the industries where we would focus our interviews. We first ranked industries using their North American Industry Classification System (NAICS) code by the number of SDVOSBs that were CCR registrants in the industry. This yielded the industries with the greatest number of registered SDVOSBs. We next ranked industries by federal contract dollars spent in them in FY 2011. This yielded the industries with the largest amount of federal spending. We then ranked industries by the total FY 2011 dollars that went to SDVOSBs. This yielded industries where SDVOSBs were successful in winning federal contracts. We also ranked industries by the percentage of SDVOSBs that were registered in them.

We then identified the top 20 industries in each of the four industry rankings: number of SDVOSBs in the CCR, federal expenditures, federal contract dollars going to SDVOSBs, and percentage of SDVOSBs registered.[1] Engineering Services (NAICS number 541330) was in the top four for the first three of these rankings, and Commercial and Institutional Building Construction (NAICS number 236220) was in the top five for the first three of them. Facilities Support Services (NAICS number 561210) was in the top 20 of all four rankings. Facilities Support Services was also among the industries in which women-owned small businesses were found to be underrepresented in federal contracting and policies had been put in place to address this underrepresentation.[2] No other industries were as close to meeting our selection criteria as these three.

Interview Methodology

Strategy for Sampling SDVOSBs and Contract Staff in the Chosen Industries

After identifying the three industries on which to focus, we developed a sampling plan. To avoid conflating SDVOSB barriers with barriers faced by other types of DoD suppliers (e.g., non-traditional suppliers, small businesses, and businesses in other preference programs), we sampled from SDVOSBs with a broad range of characteristics. We did this by organizing the SDVOSBs into several groups and then sampling from all groups. First, we identified SDVOSBs with various sources of federal funding. We selected firms that (1) had no federal contracts (but that had registered to do business with the federal government), (2) had funding only from DoD, (3) had funding from other federal agencies but not DoD, or (4) had funding from both DoD and other federal agencies.[3] We randomly sampled five firms from each of these funding categories. Second, we recognized that nearly all of the firms sampled in the funding categories were SDVOSBs only; they were not at the same time registered in other preference programs that might face different barriers. We therefore further identified SDVOSBs by whether they qualified for other preference programs as well. These included those that were (5) women-owned, (6) Native American individual-owned, (7) Native American group-owned, (8) minority-owned, (9) small disadvantaged businesses, and (10) were Small Business Administration (SBA) certified. In each of these categories, we randomly selected four additional firms. We sampled only four firms (compared with the five in the funding categories) because we sought to include the experience of SDVOSBs that were also in another preference program; we did not need a large sample in any one particular program. In total, we thus identified 10 groups of SDVOSBs within each of the three industries, yielding 44 SDVOSBs sampled in each NAICS category, for a total of 132 companies.[4]

---

[1] Seven industries were in the top 20 of the first three rankings. However, some industries were more often than others in the top five. The last ranking by percentage of SDVOSBs registered had only four industries that overlapped with any of the top 20 of the first three rankings. Further, three of the four industries overlapped with the top 20 ranking only for federal contract dollars going to SDVOSBs. Many of the industries in the percentage of SDVOSBs registered had few total contractors registered and relatively low numbers of federal dollars and total SDVOSBs registered.

[2] In 2010, the Small Business Administration identified 83 industries in which women-owned small businesses are underrepresented or substantially underrepresented in the federal contract marketplace. See U.S. Small Business Administration, 2010.

[3] This distinction between DoD and other federal agencies was requested by the study's sponsor, DoD's OSBP, to ensure that barriers faced by DoD contractors would be included.

[4] Note that a company could cover more than one category, e.g., funding from DoD only, no other preference program, and SBA-certified.

We also sampled contracting specialists who had worked on DoD contracts with SDVOSBs. We restricted this sample to contracting specialists only from DoD (and not other federal agencies), because OSBP was unable to tabulate the contact information for federal staff. We had data available on DoD staff and tabulated those data for use here. We identified all DoD contracts with SDVOSBs in each of the three industries. We then identified the contracting specialists associated with each of these contracts and randomly selected 20 from each industry.

Interview Operations

Once the sample was determined, we planned for several other issues relevant to interview response rates and operations. For both SDVOSBs and contract staff, we addressed potential concerns about responding to the interviews. These included legitimacy of the request, time constraints, and anonymity concerns. We attached a letter of support from the study's sponsor (the director of OSBP) and a summary of the project to assure potential interviewees that the request was legitimate. The letter of support also served to give the study a high priority. All participation was voluntary, and we addressed potential time constraints by developing an SDVOSB-interview protocol that could be completed within 45 minutes and a contract-staff interview protocol that could be completed within 15 minutes. We included this time information in our interview requests and scheduled only this amount of time from those who responded. (A few SDVOSB interviews and one contracting staff interview lasted up to 15 minutes longer than the scheduled time, and we confirmed with those interviewees that they were willing to speak for the additional time.) Finally, to address potential anonymity concerns, we assured all potential interviewees that we would not identify them by name or discuss in our findings other characteristics (e.g., participation in another preference program) that might identify them. We explained that we would aggregate our findings across all interviewees and describe barriers on a level that could generalize to all SDVOSBs. We also assured interviewees that we were not recording their responses but only taking written notes.

We contacted 132 firms to interview, because our experience soliciting private firms for prior government studies was that private firms were reluctant to respond, indicating that we would likely obtain a low response rate. Thirteen companies responded to our initial request for an interview, which is consistent with response rates of comparable studies (see, for example, Cox, Moore, and Grammich, 2013; Moore et al., 2011). Because these interviews included a broad range of characteristics and ample repetition in responses, we were confident that we had heard many key themes faced by SDVOSBs and did not re-contact the remaining companies sampled for interviews. This does not mean that the results can be considered a statistically representative sample of SDVOSBs. Rather, as qualitative data, they provide insight that quantitative, statistically generalizable data cannot. In this case, the in-depth interviews provide details about the barriers experienced by at least some SDVOSBs in the processes of bidding and contracting with the federal government. It is possible that the firms we interviewed constituted a biased sample, because those who responded to our interview request might be those with an agenda (for example, those who had faced the most barriers). At the same time, we believe that the findings presented here lend key insight into barriers faced by many SDVOSBs, because the interviewed firms represented a wide range of characteristics, including all but one of the 10 sampling groups identified above, and because we heard ample repetition of barriers and themes in the interviews.

In addition to these solicited responses, we also received eight requests to participate in the study from SDVOSBs we had not contacted but that had heard about the study. We invited these firms to submit written responses to our interview questions. Although all had initially said they would respond, only two did so, and those responses did not include any new barriers that we had not already learned about.

In contrast to the participation by SDVOSBs, none of the 60 contracting specialists we contacted responded to our initial request for an interview. After repeated follow-ups, we obtained responses to our questions from five staff members—four written and one by interview. These responses did include some repetition, which suggested that they reflected the experience of contracting staff.

There are two possibilities for the relatively low response rates of contracting staff. One is that the timing of the request corresponded with a particularly busy time of year for them (it was near the end of the fiscal year, when they tend to be most busy). This was unavoidable, given when we received necessary background information from OSBP. The other, potentially overlapping possibility is that contracting staff members are not disposed to respond to requests, regardless of the source of the request or the timing. This was a problem that many of the SDVOSBs we interviewed reported; as we explain below, many of the firms had repeated problems getting responses from contracting staff. Other research on non-traditional suppliers to DoD (i.e., firms that have not done a lot of DoD business) have reported similar problems in getting responses to requests for information from DoD contracting staff (Cox, Moore, and Grammich, 2013).

The interview questions derived from a literature and policy review summarized in a background paper prepared by OSBP. We identified themes from the policy history and the relevant studies in that paper. These included employment characteristics, common industries, legislation and policies affecting SDVOSBs, and potential barriers. We then translated these themes into questions. We also added some questions referring to barriers to non-traditional suppliers—those seeking to work with the federal government but that have won only small contracts or no contracts. The low rate of federal contracting with SDVOSBs means that, by definition, many SDVOSBs are non-traditional suppliers as well. Including the questions about barriers to non-traditional suppliers enabled us to ascertain whether barriers that SDVOSBs report were actually SDVOSB issues or were also non-traditional-supplier issues. Once we had a protocol of questions, we vetted the questions with colleagues for accuracy and clarity. Both the SDVOSB and contracting staff protocols can be found in Appendix B.

Our SDVOSB interviews lasted 45 to 60 minutes each and were attended by two researchers. To maintain accuracy in the responses given, both researchers took notes, but one focused on note-taking while the other asked the interview questions. Our contract staff responses lasted 15-30 minutes or were submitted in written form.

Analyzing the responses involved a multistage process. First, we combined notes for each interview. We then looked across the interviews at responses to each question and identified possible themes. Next, we looked across questions to identify additional possible themes. We then analyzed our themes for overlapping or overarching issues. We followed this procedure first for the SDVOSB interviews and then for the contracting staff interviews. We then compared and combined the results from the two groups.

Interview Results

The SDVOSBs interviewed were similar to those sampled with regard to their qualification for other preference programs, their size, and, to a lesser degree, their contracting. Table 2.1 shows the average contract sizes and the size of the firms (in terms of annual revenue and number of employees) for those sampled and for those interviewed. The firms interviewed had received more funding, on average, than those sampled, which suggests that, on average, the interviewed firms had either faced fewer barriers or had been more successful in overcoming barriers than the firms sampled. This could serve as a counterweight to the possibility noted above, that the interviewed firms are more likely than those sampled to have faced more barriers. The annual revenues of the firms sampled and interviewed were more similar ($5.4 million and $4.3 million, respectively), although the upper end of the revenue range was notably lower for the firms interviewed ($28 million) than for those sampled ($151 million). Finally, the average number of employees was similar for the firms sampled and those interviewed (38 and 30, respectively), although, again, the upper end of the range was substantially lower for the firms interviewed (132) than for the firms sampled (632).

The interviewed firms also represented a broad range of other characteristics. In terms of business experience, these characteristics included years in business, ownership of other businesses, and winning commercial as well as federal contracts. On average, the firms had been in business 10 years, but they ranged from relative newcomers with three years in business to firms with 25 years of experience. Five of the interviewees owned other firms and eight owned only their own business. Most of the firms focused on either federal or commercial contracts as the bulk of their business. Eight had mostly federal contracts, reporting that 10 percent or less of their business was commercial; four had mostly commercial contracts, reporting that 80 percent or more of their business was in the private sector. One firm reported that its business was split about 50-50 between federal and commercial contracts.

In terms of federal contracting experience, the companies also covered a broad range. All had bid on at least some federal contracts. The lowest reported number of bids was 15; the highest was approximately 1,250. Only part of this wide range was because some companies

Table 2.1
Characteristics of SDVOSBs Sampled and SDVOSBs Interviewed

	Firms Sampled	Firms Interviewed
Contract amount		
Median	$901,591	$2,171,795
Mean	$4,596,670	$6,733,901
Range	$2,501 to $36,413,179	$32,997 to $26,081,914
Annual revenue		
Median	$5,368,923	$4,285,214
Range	$0 to $150,792,323	$25,000 to $28,000,000
Number of employees		
Mean	38	30
Range	0 to 632	3 to 132

had been bidding for more years; in any given year, the number of reported bids submitted by these companies ranged from two to more than 50. They also varied in how many federal contracts they had won, from zero to more than 50 awards.

Interviewees also had experience working with many different federal agencies. These included

> Bureau of Engraving and Printing
> Bureau of Land Management
> Customs and Border Protection
> Department of Agriculture
> Department of Defense
> Department of Energy
> Department of Homeland Security
> Department of Interior
> Department of Justice
> Department of Labor
> Department of State
> Department of Veterans' Affairs
> Environmental Protection Agency
> Federal Bureau of Investigation
> Federal Emergency Management Agency
> Federal Maritime Commission
> General Services Administration
> National Aeronautic and Space Administration
> National Institute of Standards and Technology
> U.S. Geological Survey
> U.S. Marshals Service.

The most common among these agencies were the VA and DoD. Multiple SDVOSBs reported multiple contracts with each of these agencies. The SDVOSBs interviewed reported the most difficulties in working with the VA. They also reported the most work with the VA, which makes it impossible to tell whether the problems reported represent actual barriers unique to the VA or are simply the result of more interactions with the agency. These problems included a VA certification process for SDVOSBs that was lengthy, prone to errors, and difficult to correct when errors occurred. Interviewees also described scopes of work in requests for proposals (RFPs) that were outdated, were inaccurate, or appeared to be recycled from earlier projects. They described excessive safety requirements (e.g., for construction contracts) that extended to office workers who would not face safety risks. They described VA management of contracts as being inflexible and idiosyncratic, once bids were awarded. Not all interviewees who worked with the VA described all of these problems, but most described at least some of them.

Barriers Faced Particularly by SDVOSBs

Across all agencies, the interview results identified three barriers unique to these firms because they were SDVOSBs. These are the way that the SDVOSB contracting program is defined as a goal and not as an obligation, a lack of understanding among many of the SDVOSBs about the program, and a barrier to receiving subcontracts in the award stage that had been planned in the bidding stage.

The definition of the SDVOSB contracting program poses the first barrier these interviewees described. When first instituted, the specific language in the Federal Acquisition Regulation (FAR) reads, "The contracting officer *may* set-aside acquisitions exceeding the micro-purchase threshold[5] for competition restricted to service-disabled veteran-owned small business concerns" (Federal Acquisition Regulation, 2012b, emphasis added). In contrast, other preference programs are presented as a requirement. For example, the language in the FAR for the HUB-Zone program reads, "A participating agency contracting officer *shall* set aside acquisitions exceeding the simplified acquisition threshold[6] for competition restricted to HUBZone small business concerns" (Federal Acquisition Regulation, 2012a, emphasis added). This distinction recently changed as a result of the 2010 Small Business Jobs Act, which established parity among the various programs (U.S. Congress, 2010). When incorporating this change into the FAR, program administrators noted their intention to clarify "that there is no order of precedence among the small business socioeconomic contracting programs" and that the SDVOSB contracting program had equal precedence with HUBZone and women-owned small-business programs (*Federal Register*, 2012).

Still, this distinction of "may" versus "shall" has meant that federal procedures for SDVOSB procurement have been weaker than for other programs. This has had significant consequences. Most important, the language of the regulation meant that contracting staff had greater discretion on whether to set a contract aside for only SDVOSBs to apply[7] and whether to award a contract to an SDVOSB. Many of the SDVOSBs and some of the contract staff we interviewed maintained that this discretion has resulted in a lower priority being placed on SDVOSB contracting. Even with the recent change that establishes parity among the programs, this previously lower priority remains aggravated by the fact that several other preference programs have higher goals associated with them. Setting aside an RFP for another program helps contract staff meet that other program's higher goal. Not all contracting staff with whom we spoke viewed the SDVOSB program as having a lower priority; some reported that they treated it the same as any preference program.

Within this context, the SDVOSBs we interviewed described how they are often not in a position to command a higher priority. They can be less-attractive suppliers, because they often lack knowledge of the FAR and of the bidding process, which means that contracting staff and customers have to do more work and absorb more risk to work with them. Moreover, the firms and the staff interviewed described how SDVOSBs also often lack prior experience with contracting officers and customers, which makes them relatively less known than an incumbent supplier and therefore an even greater risk.

A second but related barrier unique to SDVOSBs heard in the interviews is that many do not understand that the federal SDVOSB procurement goals are not mandatory and are not guaranteed. We heard many misconceptions about the program in the SDVOSB interviews. For example, several expected that 3 percent of federal contracts would automatically

---

[5]   The micro-purchase threshold is currently $3,000.

[6]   The simplified acquisition threshold is currently $150,000.

[7]   To set aside a contract for SDVOSBs, a contracting officer must be reasonably assured that two or more SDVOSBs will bid.

be awarded to SDVOSBs. Others thought that an SDVOSB bid would automatically have 3 percentage points added to its score for award decisions (as our interviewees reported some state programs do). This lack of understanding of the program contributes to a sense of disillusionment among some SDVOSBs, which described the low percentage of contracts awarded to SDVOSBs as a betrayal or even as a violation of the program. This, in turn, discouraged them from bidding further.

A potential source of clarification for this problem is the federal information seminars offered to SDVOSBs about the program. Yet even though the SDVOSBs we interviewed had attended these seminars, many still had misconceptions about the program.

Finally, for prime contractors, including an SDVOSB on a bid demonstrates an effort to satisfy federal SDVOSB subcontracting goals and improves a bid's chances of winning. Nevertheless, insufficient federal oversight after contracts are awarded, stemming largely from insufficient information, fails to encourage prime contractors to include SDVOSBs in the actual work and poses a third barrier. Many SDVOSBs described experiences with prime contractors who included them as subcontractors on bids but rarely on work that was awarded. Federal contractors are supposed to report subcontracting efforts semiannually in the electronic Subcontracting Reporting System, but this system shows only dollar totals in each program area (e.g., SDVOSB, HUBZone) and not the names of contractors. In addition to these reports, and at least within DoD, bids that have more definitive subcontracting plans (e.g., that include memoranda of understanding with specific subcontracting firms) are evaluated higher. Despite these federal efforts to encourage SDVOSB subcontracting, we heard multiple examples of failure to verify whether the subcontracting actually occurred and of no consequences when it did not. This is not surprising, given that contracting staff lack the information they need. The SDVOSBs we interviewed described how they are reluctant to complain about this problem to federal officials for fear of being labeled as whistle blowers and being excluded from potential future subcontracts.

The effect of this barrier extends beyond the lost subcontracting business. Subcontracting is particularly important for businesses new to federal contracting, because many SDVOSBs, as with any non-traditional supplier, need to build relationships and establish good performance histories with federal customers and with other suppliers. Thus, a barrier from lack of federal oversight to ensure that subcontracting agreements are fulfilled serves as an indirect barrier to future prime contracts as well.

Barriers Also Faced by Non-Traditional Suppliers and Small Businesses

The interviewees also described encountering many of the same barriers that non-traditional suppliers and small businesses face as well. Perhaps the most common among these was the inherent complexities of the FAR and of the federal bidding process, both of which require thorough, detailed knowledge of the FAR as well as a need to hire employees with specialized knowledge to develop the bids. This requires a level of capital that many SDVOSBs reportedly lack.

A second, related barrier was a lack of federal educational and networking opportunities that might fill some of this void, whether by directly teaching the knowledge required or by teaming SDVOSBs with more experienced suppliers to acquire it. Many interviewees described the knowledge offered at federal SDVOSB conferences as basic and not at the advanced level needed to learn how to develop a successful bid. Some contract staff confirmed that SDVOSBs seem less knowledgeable about the FAR than some other small businesses, which they assumed had received stronger training or had more experience.

A third barrier faced by many non-traditional suppliers that SDVOSBs also described was a lack of communication from key contracting personnel. The interviewees described requirements in RFPs that were opaque or too general and points of contact for RFP questions who, when pressed, did not clarify the requirements and sometimes did not even return calls or emails. They also described little feedback on bids, whether successful or unsuccessful.

A fourth barrier described in the interviews and shared by other groups was the slow federal processes for award decisions and making final payments. Small businesses in particular have a hard time supporting staff and business overhead for the long time between bid and award in federal contracting processes. Similarly, backlogs at the Defense Contract Management Agency or the Defense Contracting Auditing Agency could hold up final payments that represented significant revenue for very small or new firms.

A fifth barrier was insufficient knowledge among the SDVOSBs interviewed about their chances of winning a bid. Some contract staff were much clearer about the amount of money being spent in a particular industry and on the number of qualified firms in those industries that might bid on work. The SDVOSBs interviewed did not have this knowledge and could not use it to make informed decisions about whether to invest in a bid.

A sixth barrier was the risk of wasting resources on developing a bid that was likely to go to an incumbent or another established supplier or even be subsequently cancelled. This is an issue for many of the SDVOSBs interviewed, because they often have more limited bid-and-proposal money and are often non-traditional suppliers and do not have established relationships with contracting offices.

Finally, the SDVOSBs that were also non-traditional suppliers are more likely to be in an emerging industry that does not fit existing NAICS categories. As such, they have a harder time being identified for relevant work in RFPs that use existing NAICS classifications of work.

To deal with these barriers, the SDVOSBs noted some limited sources of support. These included the federal training conferences (e.g., hosted by SBA or DoD) that did provide very basic information to help a firm get started. They described more detailed information and assistance coming from state-level Procurement Technical Assistance Centers (PTACs) in several locations. A few SDVOSBs described partnerships with other suppliers on bids as the most useful source of information for learning the many details involved in developing a successful federal bid. Firms that had been able to invest in their own bidding capabilities, either by partnering with other suppliers or by hiring staff with prior federal contracting knowledge, were able to further improve their chances of winning a bid. Finally, the SDVOSBs that described a highly customer-oriented approach to federal contracting had the greatest success. This approach extended beyond trying to satisfy the needs of the end-user to trying to understand and satisfy the needs of the contracting officer as well (e.g., submitting reports on time).

For the SDVOSBs that were either unaware of or unable to invest in these sources of support, encountering these barriers often led to some disillusionment with the program and with the federal government's commitment to veterans. The widely publicized 3 percent goal reportedly raised many SDVOSBs' expectations of gaining federal business, and they invested in their companies and in bids accordingly. Yet the apparent lack of high priority placed on the program by some contracting staff has led to fewer set-asides for SDVOSBs, and therefore fewer set-aside SDVOSB bidding opportunities, than many of the firms expected. In addition, a lack of accurate understanding of the program combined with a lack of knowledge (of the FAR, the bidding process, and the level of competition) and a lack of relationships have led to a low SDVOSB success rate for some SDVOSBs. Among those we interviewed, disillusionment was not present in the few cases where the SDVOSBs focused relentlessly on increasing their knowledge of the FAR and of the bidding process and catered to the needs of both the contracting officer and the customer when they did win an award.

CHAPTER THREE

Summary of Findings and Recommendations

This study combined quantitative analyses of service-disabled veterans and in-depth qualitative interviews with a small sample of SDVOSBs to identify the barriers that SDVOSBs may face in contracting with the federal government. The quantitative analyses indicated that service-disabled veterans are less likely than other veterans and non-veterans to be self-employed, even when differences in individual characteristics are controlled for. In addition, recent apparent increases in the number of service-disabled veterans appear unlikely to translate into a significant increase in the number of SDVOSBs with which the federal government can pursue contracting for goods and services.

Though not statistically generalizable, our interview findings suggest multiple reasons for the low rate of federal SDVOSB contracting, relative to the 3 percent federal goal.

- The level of priority placed on the program faces several challenges. The policy lays out goals rather than requirements, competing goals from other programs limit set-asides for SDVOSBs, and limited oversight of subcontracting goals means that prime contractors do not have to include SDVOSBs in awarded work.
- SDVOSBs, especially new ones, may also confront the barriers faced by small businesses and non-traditional suppliers.
- Although some sources of support for dealing with these barriers exist, they are limited.
- Given these barriers, many SDVOSBs we interviewed are disillusioned with the program. If widespread across SDVOSBs in general, this dillusionment may limit federal contracting rates further.

Given these reasons, as well as the quantitative findings that service-disabled veterans have lower self-employment and that the number of SDVOSBs is unlikely to increase, whether the federal government can reach its 3 percent goal for contracting with SDVOSBs is not clear. Identifying the number of SDVOSBs in the industries in which federal agencies spend, separately by agency, may shed further light on this issue, as would further analyses of service-disabled veterans' decisionmaking processes regarding self-employment. In addition, estimating the actual dollars and number of contracts that would be received by SDVOSBs if the goal were met would increase the understanding of this goal and of the potential effect of efforts to reduce barriers. Estimates of the degree to which the goal is being met suggest that at least an additional $1.5 billion in federal contracts would flow toward SDVOSBs if the goal could be met (U.S. General Services Administration, 2012; U.S. Office of Management and Budget, 2012). Finally, these findings suggest that reducing barriers facing SDVOSBs may involve at least three areas of effort: increasing the priority of the SDVOSB program, strengthening education and training opportunities, and continuing efforts to remove barriers to non-traditional suppliers and small businesses. We describe these in greater detail below.

Increase the Priority of the SDVOSB Program

The government wants to increase the priority of the SDVOSB program and has incorporated recent policy changes made by the 2010 Small Business Jobs Act into the FAR. Federal agencies now need to ensure that contracting staff and businesses are informed of these changes explicitly.

Agencies also need to inform staff of which industries have ample numbers of SDVOSBs, especially as compared with the numbers of firms in other preference programs.

A further step toward raising the priority placed on the SDVOSB program would be to give contracting officers the information, tools, and resources to review subcontracting use, to conduct those reviews, and to publish the results. Publishing the planned and actual subcontracting records of prime contractors (i.e., the planned and actual dollars awarded to SDVOSBs included in their bids and identified by their Dun and Bradstreet identifying numbers) would provide a needed incentive for prime contractors to follow through on subcontracting plans. This should be enabled by the expanded information in the Federal Funding Accountability and Transparency Act Subaward Reporting System.

Strengthen Education and Training Opportunities

Providing SDVOSBs with the industry analyses described above could be a first step toward strengthening education and training opportunities available to SDVOSBs. This would give SDVOSBs clear information about the level of federal agency spending in the industries in which they are registered and about their competition in those industries.

Other education could include clarification that the federal SDVOSB program (calling for 3 percent of prime contracting dollars to go to SDVOSBs) is a goal, not a mandate. In addition, more advanced information and training about the FAR and the federal bidding process would be useful. This might involve the actual FAR training-course material that contracting officers use or another similar forum. It should also involve more information about the roles and responsibilities of contracting staff to ensure that SDVOSBs have reasonable expectations of them. To the degree that any of this could be done virtually, costs could be minimized for both the SDVOSBs and federal government.

Given that several SDVOSBs described useful experiences with PTAC offices, strengthening and enhancing PTAC support would appear to be another step toward improving education and training opportunities. It might even be efficient to use them as the channels for providing more advanced information and training about the FAR and the bidding process.

Networking opportunities in any of these forums could help SDVOSBs needing to build their knowledge and their connections to identify firms with which to partner or subcontract. These relationships, in turn, can further the education and training about the bidding process.

Finally, education and training opportunities would also be strengthened by clarifying federal expectations of SDVOSBs. Those SDVOSBs interviewed that invested in a comprehensive understanding of the FAR and in a lengthy process of learning how to develop a successful bid saw dramatic improvements in their success rates. Similarly, the SDVOSBs interviewed that learned and met the needs of both the contracting officer and the customer received strong performance evaluations, strengthening their ability to win future bids. Transmitting these expectations to all SDVOSBs is an important part of the education and training.

Continue Efforts to Remove Barriers to Non-Traditional Suppliers and Other Small Businesses

Continuing efforts aimed at removing barriers to other non-traditional suppliers and small businesses would likely also help reduce the barriers experienced by SDVOSBs, because most are small and many appear to be non-traditional. These efforts include continuing to streamline the administrative requirements of the bidding process and removing inefficiencies in that process. Such streamlining might include creating a government-wide repository for the background material required in most bids. Further streamlining the award decision and payment procedures is particularly important to small firms. Improving communication between government and suppliers is another important effort at reducing these barriers. Part of this communication would include offering information about current incumbents' performance to help SDVOSBs evaluate their chances of winning a bid. Finally, because NAICS definitions are revised only every five years, helping to identify suppliers in emerging industries not clearly specified in the current NAICS definitions to both customers and contracting officers is important for some non-traditional suppliers.

Logistic Regression Methodology and Analyses

Logistic regression is a common analytic method for modeling a dichotomous outcome with multiple covariates. In this case, whether someone is self-employed or not is the dichotomous outcome of interest. The equation below shows the model's basic form:

$$\log it(p) = \log\left(\frac{p}{1-p}\right) = \beta_0 + \beta_1 X_1 + \ldots + \beta_n X_n + \varepsilon,$$

where p is the probability that someone will be self-employed and the X variables are individual characteristics associated with self-employment. The βs are coefficients of the X variables and represent the change in the log odds of a single unit change in the associated X variable (i.e., the logit of the probability). The ratio of $p/(1-p)$ is the ratio of the odds of being self-employed to the odds of not being self-employed; it is referred to as the *odds ratio*. Odds ratios greater than one indicate a positive relationship between self-employment and the associated independent variable. Odds ratios less than one indicate a negative relationship between self-employment and the associated independent variable. The tables that follow present logistic regression results in the form of odds ratios. Table A.1 presents two models of self-employment among all adults, Table A.2 presents two models of self-employment among adults who are in the paid labor force, and Table A.3 presents models of self-employment for non-veterans, veterans, and service-disabled veterans separately.

Table A.1
Odds Ratios from Logistic Regression of Self-Employment Among All Adults

Variable	Non-Veterans Omitted	Non-Disabled Veterans Omitted
Service-disabled veterans	0.33***	0.48***
Non-disabled veterans	0.69***	—
Non-veterans	—	1.45***
Education		
Graduate school	1.53***	1.53***
College graduate	1.40***	1.40***
Some college	1.25***	1.25*
(High school or less)	—	—
Age (experience proxy)		
Age in years	1.21***	1.21***
Age squared in years	1.00***	1.00***
Gender		
Female	0.42***	0.42***
(Male)	—	—
Race-ethnicity		
African American	0.46***	0.46***
Hispanic	0.75***	0.75***
Other race-ethnicity	0.75***	0.75***
(White)	—	—
Marital status		
Never married	0.66***	0.66***
Formerly married	0.81***	0.81***
(Married)	—	—
Sample size	99,066	99,066

SOURCE: Authors' tabulation of July 2010 CPS data.

NOTES: * $p < .05$, ** $p < .01$, ***$p < .001$.

Table A.2
Odds Ratios from Logistic Regression of Self-Employment Among All Adults in Paid Labor Force

Variable	Non-Veterans Omitted	Non-Disabled Veterans Omitted
Service-disabled veterans	0.43***	0.63***
Non-disabled veterans	0.69***	—
Non-veterans	—	1.44***
Education		
Graduate school	1.19***	1.19***
College graduate	1.15***	1.15***
Some college	1.08*	1.08*
(High school or less)	—	—
Age (experience proxy)		
Age in years	1.07***	1.07***
Age squared in years	1.00***	1.00***
Gender		
Female	0.51***	0.51***
(Male)	—	—
Race-ethnicity		
African American	0.49***	0.49***
Hispanic	0.75***	0.75***
Other race-ethnicity	0.81***	0.81***
(White)	—	—
Marital status		
Never married	0.65***	0.65***
Formerly married	0.83***	0.83***
(Married)	—	—
Sample size	66,473	66,473

SOURCE: Authors' tabulation of July 2010 CPS data.

NOTES: * p < .05, ** p < .01, ***p < .001.

Table A.3
Logistic Regression Results of Self-Employment, by Veteran and Disability Status

Variable	Non-Veterans	Veterans	Service-Disabled Veterans
Age in years	1.03***	N/A	N/A
Veteran cohort	N/A		
Before 1964		9.75***	5.73*
Vietnam era		3.50***	2.60+
Post-Vietnam era		2.03***	2.38
1990s		1.67*	2.10
(2000s)		—	—
Education			
Graduate school	1.07***	2.48***	3.03*
College graduate	1.08***	1.38*	1.43
Some college	1.02***	1.20	1.31
(High school or less)	—		
Gender			
Female	0.51***	0.54**	1.01
(Male)	—	—	—
Race-ethnicity			
African American	0.53***	0.61*	0.62
Hispanic	0.76***	0.86	1.12
Other race-ethnicity	0.84***	0.39*	0.01
(White)	—	—	—
Marital status			
Never married	0.64***	1.00	0.62
Formerly married	0.86***	0.82	2.71**
(Married)	—	—	—
Sample size	63,246	4,234	560

SOURCE: Authors' tabulation of July 2010 CPS data.
NOTES: + p < .10, * p < .05, ** p < .01, ***p < .001.

Interview Protocols

Business Interview Protocol for RAND Study,
"Improving Federal and Department of Defense (DoD) Prime Contract Support
of Service-Disabled Veteran Owned Businesses"
Sponsored by Office of Small Business Programs
Office of the Undersecretary of Defense, Department of Defense

We have been asked by the Department of Defense Office of Small Business Programs to analyze barriers to boosting prime contracting dollars spent with service-disabled veteran owned small businesses (SDVOSBs) within the federal government. The purpose of this interview is to identify any barriers SDVOSBs face in successfully obtaining federal contracts. We are particularly interested in barriers directly related to being a service-disabled veteran-owned small business. We will not reveal the names of persons or companies we interview. Rather, we plan to aggregate our findings across all of our interviews and characterize our findings in a general way. Your participation in this study is voluntary, and you may choose not to answer any or all of the questions that follow.

Applying for Federal Contracts

1. Please describe your experience with the process of applying for federal contracts.
    a. Have you applied for any contracts? If so, how many?
    b. How did you learn the application process?
    c. Were you able to find contracts to bid on?
    d. Were you able to get the information you needed for the application?
    e. Were you able to get answers to questions that you had?
    f. If yes to any of the above, where did you find these resources?
    g. Did you attend any DoD seminars or workshops? If yes, were these useful? Please explain.
    h. Did you receive any contracts? If yes:
        i. How many?
        ii. For what length of time?
        iii. For what federal departments?
    i. Were these all prime contracts of your own or were some of them subcontracts on another firm's contract?

2. Do you have private sector contracts?
   a. If yes, about how much of your business is in private contracts?

3. Do you believe or have you experienced that winning contracts with the federal government is easier or harder than in the commercial sector? Please explain.
   [Prompt:] Is this because:
   a. There are fewer opportunities in the federal government for the good or service that your firm offers.
   b. Competition is greater or lesser? Please explain.
   c. Small business goals for other types of small or disadvantaged businesses create greater competition. Please explain.
   d. Is any of this related to being a service-disabled veteran-owned small business?

4. Has the time or money required to apply for a federal contract or for a contract in a specific federal department made you less likely to apply? If yes, please describe.
   [Prompt:]
   a. Is any of this related to being a service-disabled veteran-owned small business?

5. Are there risks involved with federal contracts or for a contract in a specific federal department that have made you less likely to apply? If yes, please describe.
   [Prompt:]
   a. Is any of this related to being a service-disabled veteran-owned small business?

Fulfilling the Contract

6. If you have received contracts, please describe your experience working with the federal government during the course of the contract(s). Were there things the federal government or specific federal departments required or did that made it easier or harder to fulfill the contract?

Services

7. Have you received any guidance or support from veteran service organizations with federal contracts? If yes, please describe what you received and how it may have helped you.

8. Have you found that the federal government's adaptability services are sufficient for you?

Conclusion

9. Are there any other things that make contracting with the federal government difficult for you as a service-disabled veteran small business owner?
10. If the federal government could change one thing that would make its business more attractive to you, what would that be?

Company Background

11. How long have you owned your business? Is this your first/only business?
12. Does your business qualify for any other commercial sector or government socioeconomic goals?

    [Prompt:] For example, women-owned, Native American owned, located in a HUBZone

**Department of Defense Staff Interview Protocol for RAND Study,
"Improving Federal and Department of Defense (DoD) Prime Contract Support
of Service-Disabled Veteran Owned Businesses"
Sponsored by Office of Small Business Programs
Office of the Undersecretary of Defense, Department of Defense**

We have been asked by the Department of Defense Office of Small Business Programs to analyze barriers to boosting prime contracting dollars spent with service-disabled veteran owned small businesses (SDVOSBs) within the federal government. This interview is part of that study, and its purpose is to identify any barriers SDVOSBs face in successfully obtaining federal contracts. We are particularly interested in barriers directly related to being a service-disabled veteran-owned small business. We will not reveal the names of persons or offices we interview. Rather, we plan to aggregate our findings across all questionnaires and characterize our findings in a general way. Your participation in this study is voluntary, and you may choose not to answer any or all of the questions that follow.

1. What is your job title, and how long have you worked in this job?
    a. How long have you worked in federal contracting?

2. Have you worked with SDVOSBs before? If yes, approximately how many contracts?

3. Does your experience with SDVOSB contracting differ from other preference programs? If so, how does it differ?
    a. Are the requirements different for the vendors in the SDVOSB procurement program than for vendors in other preference programs? If so, please explain.
    b. Are the requirements different for you? If so, please explain.

4. For specific solicitations, are there any practices for reaching out to SDVOSBs and encouraging them to bid? Please describe.

5. Is there any mechanism for ensuring that SDVOSBs included as subcontractors on a bid are actually included in the work if the prime contractor is awarded a contract?

6. In your experience, have you observed any barriers that SDVOSBs face in trying to win federal contracts? If so, please explain.
    a. Do any of these barriers differ from those other small businesses or new suppliers face?

Bibliography

Brewin, Bob, "VA's Claims Backlog Continues to Push 900,000," NextGov. September 21, 2012. As of September 28, 2012:
http://www.nextgov.com/health/2012/09/vas-claims-backlog-continues-push-900000/58293/?oref=govexec_today_nl

Bush, George W., "Service Disabled Veterans," Executive Order 13360, Washington, D.C., 2004. As of December 10, 2012:
http://www.gsa.gov/portal/content/105166

Cox, Amy G., Nancy Y. Moore, and Clifford A. Grammich, *Identifying and Eliminating Barriers Faced by Nontraditional Department of Defense Suppliers,* Santa Monica, Calif.: RAND Corporation, 2013. As of May 7, 2013:
http://www.rand.org/pubs/research_reports/RR267.html

Department of Defense, Office of the Under Secretary of Defense for Acquisition, Technology, and Logistics, "Department of Defense Service-Disabled Veteran-Owned Small Business Strategic Plan," 2007. As of March 6, 2013:
http://www.acq.osd.mil/osbp/docs/3rd_year_dod_sdvosb_strategic_plan.pdf

Department of Veterans' Affairs, "Trends in Veterans with a Service-Connected Disability: 1986–2011," Washington, D.C.: National Center for Veterans' Analysis and Statistics, 2012. As of September 25, 2012:
http://www.va.gov/vetdata/docs/QuickFacts/SCD_trends_FINAL.pdf

Fairlie, Robert W., and Alicia M. Robb, *Race and Entrepreneurial Success,* Boston, Mass.: MIT Press, 2008.

Federal Acquisition Regulation, "Small Business Programs: Historically Underutilized Business Zone (HUBZone) Program," Part 19.1305, 2012a. As of May 13, 2013:
http://www.acquisition.gov/far/current/pdf/FAR.pdf

———, "Small Business Programs: Historically Underutilized Business Zone (HUBZone) Program," Part 19.1405, 2012b. As of May 13, 2013:
http://www.acquisition.gov/far/current/pdf/FAR.pdf

Federal Procurement Data System—Next Generation, "Small Business Goaling Reports," undated. As of March 24, 2013:
http://www.fpdsng.com/fpdsng_cms/index.php/reports

Federal Register, "Federal Acquisition Regulation: Socioeconomic Program Parity," Vol. 77, No. 42, March 2, 2012, pp. 12930–12933.

Moore, Nancy Y., Amy G. Cox, Lloyd S. Dixon, Clifford A. Grammich, Judith D. Mele, and Aaron Kofner, "Improving the Methodology for Setting Small-Business Size Standards," Santa Monica, Calif.: RAND Corporation, unpublished research, 2011.

Teachman, Jay, "A Note on Disappearing Veterans: 1980 to 2010," *Armed Forces and Society,* January 15, 2013: As of March 24, 2013:
http://afs.sagepub.com/content/early/2013/01/04/0095327X12468731.abstract

U.S. Bureau of Labor Statistics, Current Population Survey, Washington, D.C.: Division of Labor Force Statistics, July 2010. As of September 25, 2012:
http://www.bls.gov/cps/

U.S. Census Bureau, *Survey of Business Owners*, Washington, D.C.: Company Statistics Division, 2007. As of September 25, 2012:
http://www.census.gov/econ/sbo/

U.S. Congress, "Small Business Jobs Act of 2010," Public Law 111-240, Washington, D.C., 2010. As of December 10, 2012:
http://www.sba.gov/sites/default/files/files/pl106-50.pdf

———, "Veterans Entrepreneurship and Small Business Development Act of 1999," Public Law 106-50, Washington, D.C., 1999. As of December 10, 2012:
http://www.sba.gov/sites/default/files/files/pl106-50.pdf

———, "Veterans Benefit Act of 2003," Public Law 108-83, Washington, D.C., 2003. As of December 10, 2012:
http://www.acq.osd.mil/osbp/docs/pl108-183.pdf

U.S. General Services Administration, "Small Business Goaling Report," 2012. As of March 1, 2013:
https://www.fpds.gov/downloads/top_requests/FPDSNG_SB_Goaling_FY_2011.pdf

U.S. Office of Management and Budget, Website, 2012. As of March 1, 2013:
http://www.usaspending.gov/

U.S. Small Business Administration, "SBA Releases Final Women-Owned Small Business Rule to Expand Access to Federal Contracting Opportunities," October 4, 2010. As of May 7, 2013:
http://www.sba.gov/about-sba-services/7367/5524

———, "SBA Resources for Veterans," 2012a. As of March 8, 2013:
http://www.sba.gov/sba-direct/article/3627

———, "Service-Disabled Veteran-Owned Businesses," 2012b. As of May 10, 2013:
http://www.sba.gov/content/service-disabled-veteran-owned-small-business-concerns-sdvosbc

———, "Small Business Procurement–FINAL FY12/13 Goals," 2012c. As of March 24, 2013:
http://www.sba.gov/sites/default/files/files/FINAL%20FY%2012-13%20Small%20Business%20Procurement%20Goals_121611(3).pdf

Made in the USA
Coppell, TX
14 May 2022

77775328R00031